About the book

There are 4 IGCSE Mathematics higher papers in this book. These are 2 sets of papers 1, & 2 written as practice papers for IGCSE Mathematics Higher Examination. Papers are mainly focusing on Edexcel examinations as well as other similar examination boards.

These papers are written according to the new grade 9-1 syllabus and questions are potential questions for the GCSE examination. This book is written for those students who are aiming for grades 7, 8 & 9.

All the questions in this book are written by the author and they are new questions written purely to help and prepare the students to test themselves for the upcoming mathematics IGCSE higher exams.

IGCSE Mathematics higher level papers for grades 7, 8, & 9

for grade 9 to 1 syllabus by Edexcel

By Dilan Wimalasena

IGCSE Mathematics higher level papers for grades 7, 8 & 9

IGCSE Mathematics higher level papers for grades 7, 8 & 9

Contents

	Page
Section A	7
Paper A1	9
Paper A2	19
Section B	29
Paper B1	31
Paper B2	41

IGCSE Mathematics higher level papers for grades 7, 8 & 9

Section A

IGCSE Mathematics higher level papers for grades 7, 8 & 9

IGCSE Mathematics higher level papers for grades 7, 8 & 9

Paper A1
Time 2 hours

IGCSE Mathematics higher level papers for grades 7, 8 & 9

1. Show that

$$\frac{4\frac{1}{2} \div (3\frac{1}{3} - 2\frac{3}{4})}{2\frac{2}{3} \times 1\frac{1}{5}} = 2\frac{23}{56}$$

(5 marks)

2. Harry drove 8km in 20 minutes and then he walked 800m in 12 minutes. Calculate the average speed of his journey.

(4 marks)

3. Simplify

$$\frac{(3x^4)^3 \times (2x^3)^2}{(6x^5)^2}$$

(5 marks)

4. Expand and simplify

$$(4x - 3)^2 - (3x + 1)(5x - 2)$$

(6 marks)

5. Rebecca has 0.65 chance of winning a game. She plays 2 games. If she wins the first game, then her probability of winning the second game is 0.8. However, if she loses the first game, then her probability of winning the second game is only 0.45.

i) Draw a tree diagram to represent all possible outcomes.

(3 marks)

ii) Calculate the probability of Rebecca winning at most one game.

(3 marks)
(total 6 marks)

6. A cylinder has radius 6cm and height 15cm. The cylinder is melted down and made into a sphere. Calculate the radius of the sphere.

(5 marks)

7. An apartment is £250,000. Its value increases by 6% every year. John says the value will be more than £600,000 after 15 years. Is John correct?

(4 marks)

8. A circle has radius 4.8cm. The circumference of the circle is 1.5 times the perimeter of a square. Calculate the area of the square.

(6 marks)

9. i) Write down two numbers with a highest common factor of 72.

(3 marks)

ii) Work out the highest common factor of 108, 144 & 180.

(4 marks)
(total 7 marks)

10. Triangle ABC is right angled with the right-angle ABC. The length AC is 20cm to one significant figure and the length BC is 8.5cm to one decimal place.

Calculate the lower bound of the length AB.

(5 marks)

11. Factorise fully

$$i)\ 6x^2 - 11x - 30$$

(3 marks)

$$ii)\ 3x^3 - 192x$$

(3 marks)

$$iii)\ 2a^2 - 9ab - 5b^2$$

(4 marks)
(total 10 marks)

12. Solve the following equations

i) $\dfrac{2-x}{3} + \dfrac{1-3x}{2} = 3$

(4 marks)

ii) $5x - 7y = 30$
    $4x - 3y = 11$

(4 marks)

iii) $x^2 - y^2 = 48$
     $x + 3y = 4$

(6 marks)
(total 14 marks)

13. A village has 74 residents. 68 of them have televisions and 57 of them have computers. 4 people do not have neither a television nor a computer.

i) Represent the above information in a Venn diagram.

(4 marks)

ii) What percentage of the village have both a television and a computer.

(2 marks)
(total 6 marks)

14. Use algebra to show that the recurring decimals
$$0.\dot{7} - 0.4\dot{1}\dot{8} = \frac{178}{495}$$

(5 marks)

15. Here are the first 4 terms of an arithmetic sequence.

$$-1, 2, 5, 8, \ldots \ldots$$

Calculate the sum of all the terms from the 31st term to the 80th term.

(5 marks)

16. A tringle has lengths 13cm, 20cm & 25cm. Calculate the area of the triangle.

(5 marks)

17. Sketch the graph of $y = \cos x$.

(2 marks)

Total for paper: 100 marks

End

IGCSE Mathematics higher level papers for grades 7, 8 & 9

IGCSE Mathematics higher level papers for grades 7, 8 & 9

Paper A2
Time 2 hours

1. The table shows information about marks of 32 students.

Marks	Frequency	
20-50	6	
50-60	8	
60-80	9	
80-90	6	
90-100	3	

i) Draw a cumulative frequency curve for the information.

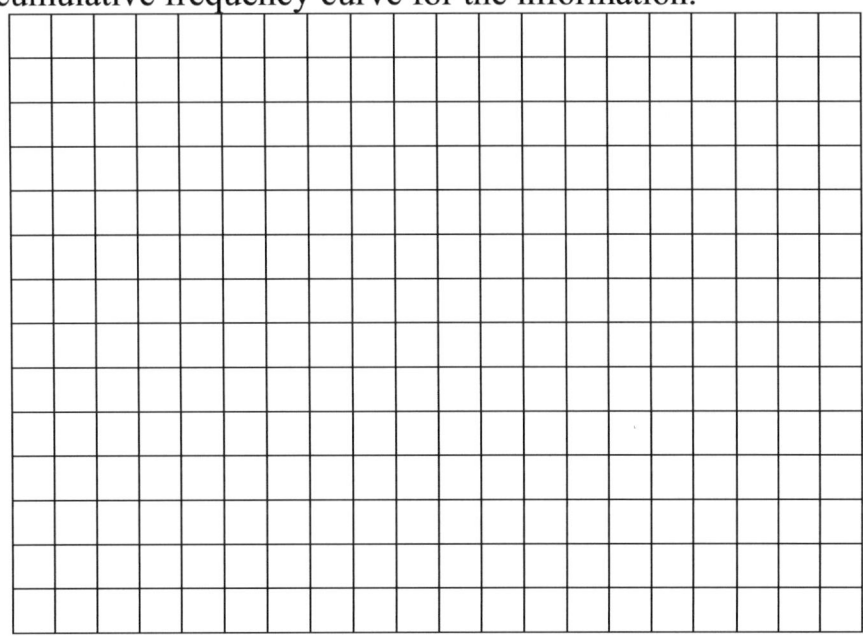

(2 marks)

ii) Work out the median mark.

(2 marks)

iii) Work out the interquartile range.

(3 marks)

iv) Calculate the percentage of students with more than 85 marks.

(3 marks)
(total 10 marks)

2. Use rule and compass to construct an angle of 105°. You must show all your construction lines.

(5 marks)

3. There are 40 people in a group. 20 of them like cricket, 18 like rugby and 25 like hockey.

4 like all three.
6 people like cricket and rugby only.
11 like cricket and hockey.
9 like rugby and hockey.

i) Draw a Venn diagram to represent the above information.

(4 marks)

ii) What is the probability of someone liking exactly 2 out of the 3 sports.

(2 marks)
(total 6 marks)

4. A cylinder has radius 3cm and height 8cm. It has the same volume as a sphere. Calculate the radius of the sphere.

(4 marks)

5. Andrew, Beth and Celia shared £$(40x + 60)$ in the ratio $5: x: 9$. Celia has £60 more than Andrew. Work out the value of x.

(5 marks)

6. i) Convert 90 km/h into m/s.

(3 marks)

ii) Convert 2500 cm^2 into m^2.

(3 marks)
(total 6 marks)

7. Solve the simultaneous equations.
$$xy = -20$$
$$x - 3y = 19$$

(5 marks)

8. A circle has centre $C(-1, 2)$ and passes through the point $P(5, 8)$.

i) Work out the equation of the tangent to the circle at the point P.

(4 marks)

ii) The tangent above meets the x axis at A and y axis at B. Work out the area of the triangle OAB where O is the origin.

(5 marks)
(total 9 marks)

9. A sector has radius 5cm and perimeter 13.5cm. Calculate the area of the sector.

(6 marks)

10. Make x the subject of the following formulae

i) $y = \sqrt{\dfrac{5-2x}{3+7x}}$

(4 marks)

ii) $\dfrac{1}{y} = \dfrac{2}{x} - \dfrac{a}{b}$

(4 marks)
(total 8 marks)

11. Show that

$$\frac{\frac{5}{\sqrt{3}} - 1}{\frac{2}{\sqrt{3}} + 1} = 13 - 7\sqrt{3}$$

(5 marks)

12. y is inversely proportional to x^2. When $x = -9, y = 4$.

Work out the values of
i) y when $x = 3$.

(4 marks)

ii) x when $y = 9$.

(4 marks)
(total 8 marks)

13. A solid cuboid has length $(3x+1)cm$, width $(x+1)cm$ and height xcm. It has a total surface area of $164cm^2$. Work out the value of x.

(5 marks)

14.

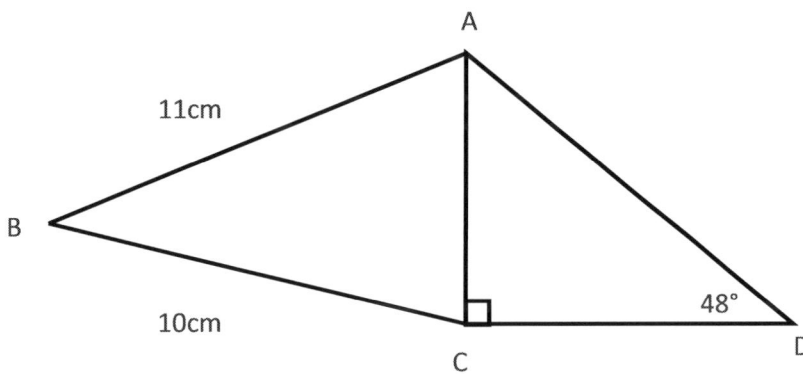

The area of the triangle ABC is $23cm^2$. Calculate the area of the triangle ACD.

(6 marks)

15. Simplify the following fully

$$\frac{(3x-1)(4x+5)}{12x^2-4x} \div \frac{16x^2-25}{x^3-x}$$

(5 marks)

16. The functions f & g are such that
$$f(x) = 4x+1, \qquad g(x) = \frac{5x-2}{3}$$

i) Find $f^{-1}(4)$

(3 marks)

ii) Solve the equation
$$fg(x) = f^{-1}(4)$$

(4 marks)
(total 7 marks)

Total for paper: 100 marks

End

Section B

IGCSE Mathematics higher level papers for grades 7, 8 & 9

Paper B1
Time 2 hours

1. Solve the following equations.

i) $\dfrac{5}{2x+7} = \dfrac{3}{4x-1}$

(3 marks)

ii) $\dfrac{4}{x-3} - \dfrac{11}{2x+1} = 1$

(5 marks)
(total 8 marks)

2. Work out the length of PR. $PQ = (2+\sqrt{2})\,cm$ & $QR = (2-\sqrt{2})\,cm$.

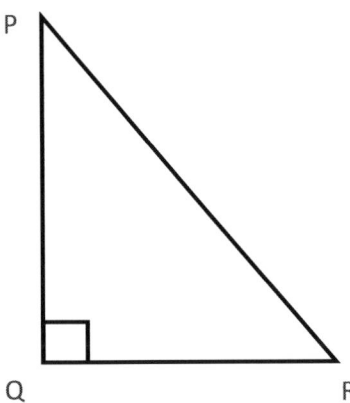

(4 marks)

3. The triangle ABC has vertices $A(1,2), B(3,4)$ & $C(3,2)$.

i) Plot the triangle ABC on the grid below.

(1 mark)

ii) Reflect the triangle ABC to the triangle PQR on the line $x = -1$.

(3 marks)

iii) Enlarge the triangle PQR by scale factor -2 about the point (-1,0)

(3 marks)

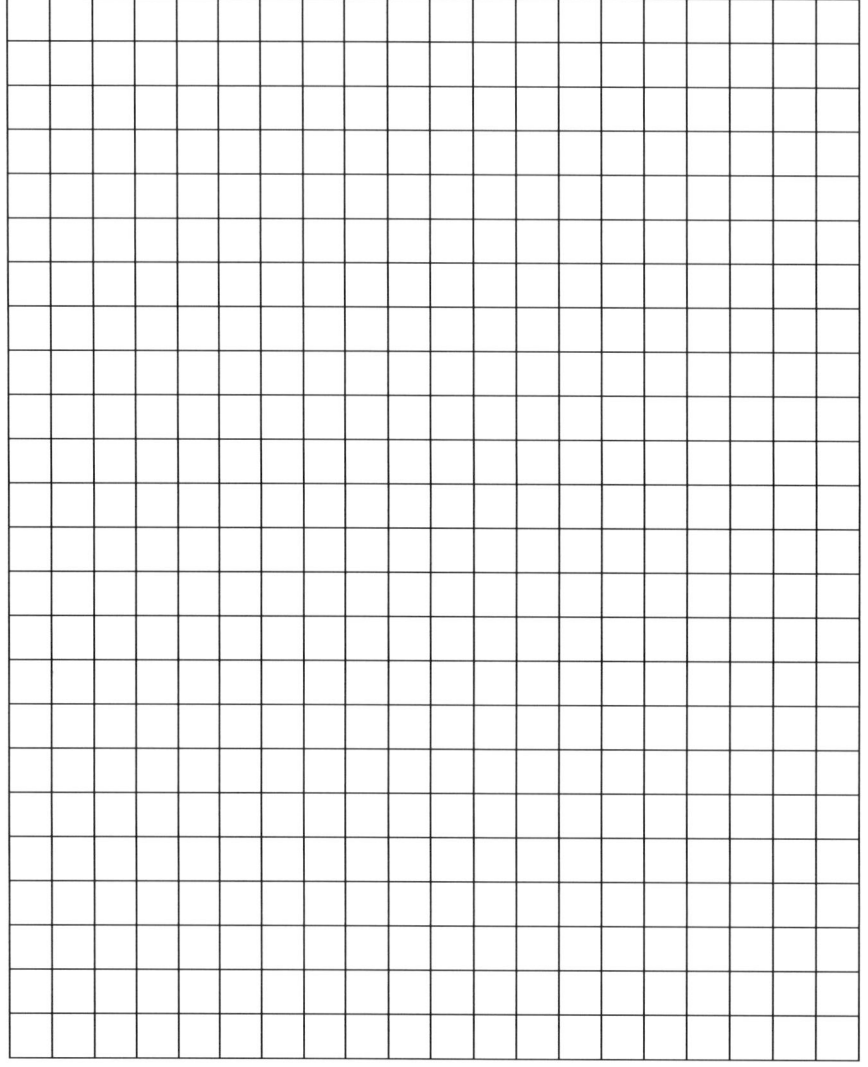

(total marks 7)

4. Work out the values of a, b & c.
$$3x^2 - 12x - 7 = a(x - b)^2 + c$$

(5 marks)

5. Here are the first 5 terms of an arithmetic sequence.
$$-5, -1, 3, 7, 11, \ldots \ldots$$

i) Work out the 45th term of the sequence.

(2 marks)

ii) Work out the sum of the first 45 terms of the sequence.

(4 marks)
(total 6 marks)

6. A particle P travels along a straight line through a point O so that at time t seconds after passing through O its displacement from O is xm, where
$$x = t^3 - 3t^2 + 5t - 7$$

i) Calculate the initial velocity of the particle.

(4 marks)

ii) Calculate the time when the acceleration is zero.

(4 marks)
(total 8 marks)

7. George bought an apartment in London for £350,000 for investment. The value of the property increases by 6% per year. The apartment receives a rental income of 4.5% a year. George decided to rent out the apartment as soon as he bought it and sold it after 3 years. Calculate his total pre-tax profit from the investment.

(5 marks)

8. There are some sweets in a bag. All the sweets are either red, yellow, blue, or green. Total number of red and yellow sweets are double the total number of blue and green sweets. The number of red sweets is $\frac{1}{4}$ of the number of yellow sweets. What percentage of the total sweets are yellow.

(6 marks)

9. The vector $\overrightarrow{OD} = k(3b + 10c)$. Work out the value of k.

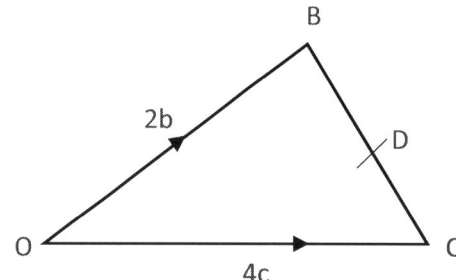

(5 marks)

10. Solve the following simultaneous equations graphically.
$$y = x^2 - 3x - 3$$
$$y = 2x - 1$$

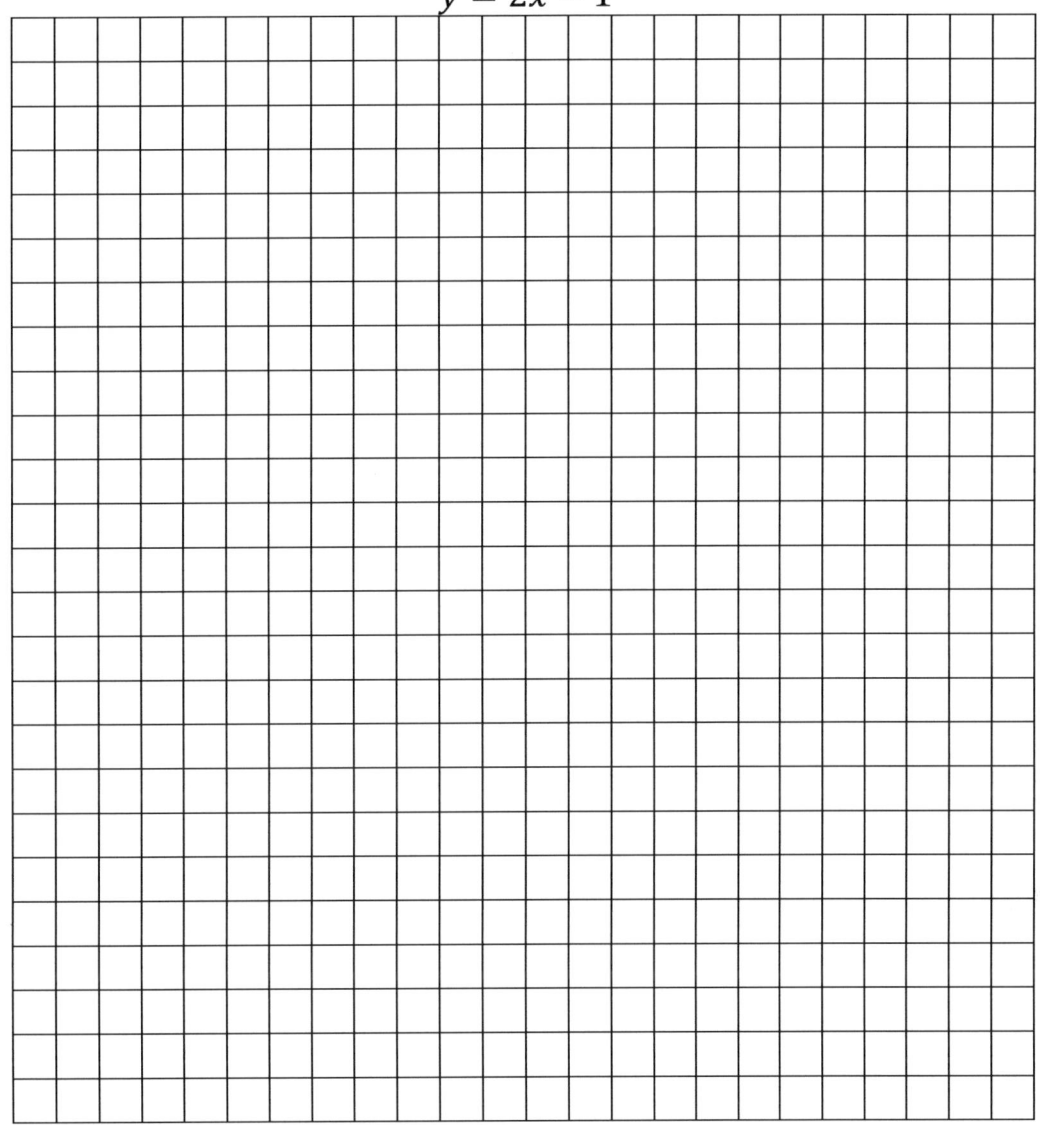

(6 marks)

11. $x = 1.25 \times 10^7, y = 3.5 \times 10^{-5}, z = 2.4 \times 10^9$
Work out the value of
$$(x + y)^2 - yz$$
Write your answer in standard form.

(4 marks)

12. Calculate the areas of the following triangles

i)

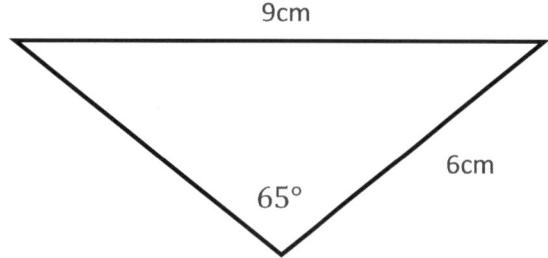

(4 marks)

ii)

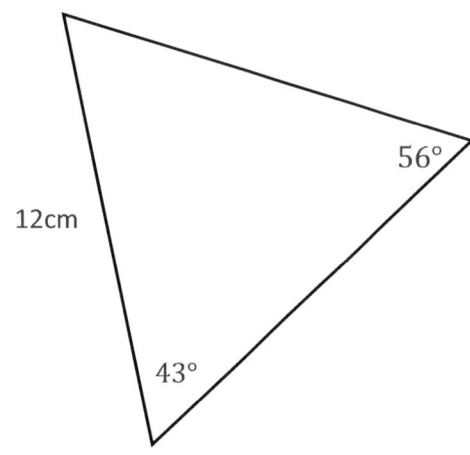

(4 marks)
(total 8 marks)

13. The curve has equation $y = x^3 - 2x^2 + x - 5$.

i) Find $\frac{dy}{dx}$

(3 marks)

ii) Find the range of values of x for which the curve C has a positive gradient.

(5 marks)
(total 8 marks)

14. 30 people in a sixth form have been surveyed. 24 of them study Mathematics and 15 study English. 2 people do not study neither Mathematics nor English.

i) Represent the above information in a Venn diagram.

(3 marks)

ii) Work out the probability of
a) a person studying only Mathematics

(2 marks)

b) both Mathematics and English.

(2 marks)
(total 7 marks)

15. A cone has radius 5cm and vertical height 12cm. It has the same surface area as a cylinder with height 8cm. Calculate the radius of the cylinder.

(6 marks)

16. The functions f & g are such that
$$f(x) = 4x - 5, \qquad g(x) = 2 - 3x$$

i) Work out

a) $fg(-1)$

(2 marks)

b) $g^{-1}(x)$

(2 marks)

ii) Hence, solve the equation
$$g^{-1}(x) = fg(-1)$$

(3 marks)
(total 7 marks)

Total for paper: 100 marks

End

Paper B2
Time 2 hours

IGCSE Mathematics higher level papers for grades 7, 8 & 9

1. Work out the highest common factor of 54, 72, 108, & 126.

(4 marks)

2. Here are the first 4 terms of a quadratic sequence.
$$-1, 5, 13, 23, ...$$

Work out an expression for the nth term of the sequence.

(4 marks)

3. A school hall floor area is 50m long and 40m wide. The hall is it be tiles using tiles which are 80cm long and 50cm wide. Each tile cost 18p. The builder charges £135 per 250 square metres.

Calculate the total cost of tiling.

(5 marks)

4. A brand new car is £24,000. The value of the car depreciates by 25% during the first 2 years and by 12% after that. Calculate the value of the car at the end of the 5th year.

(4 marks)

5.
Liquid X	Liquid Y
Density $2.5 g/cm^3$	Density $1.8 g/cm^3$

$200 ml$ of liquid X is mixed with $300 ml$ of liquid Y to create the liquid Z.

Calculate the density of liquid Z.

(6 marks)

6. Solve the simultaneous equations
$x^2 - y^2 = 85$
$2y + x = -1$

(6 marks)

7. A (-3, -1), B (7,5). Work out the equation of the perpendicular bisector to the line AB.

(5 marks)

8. Draw a histogram to represent the following.

Salaries	Frequency		
200-300	6		
300-500	10		
500-800	3		
800-1000	2		

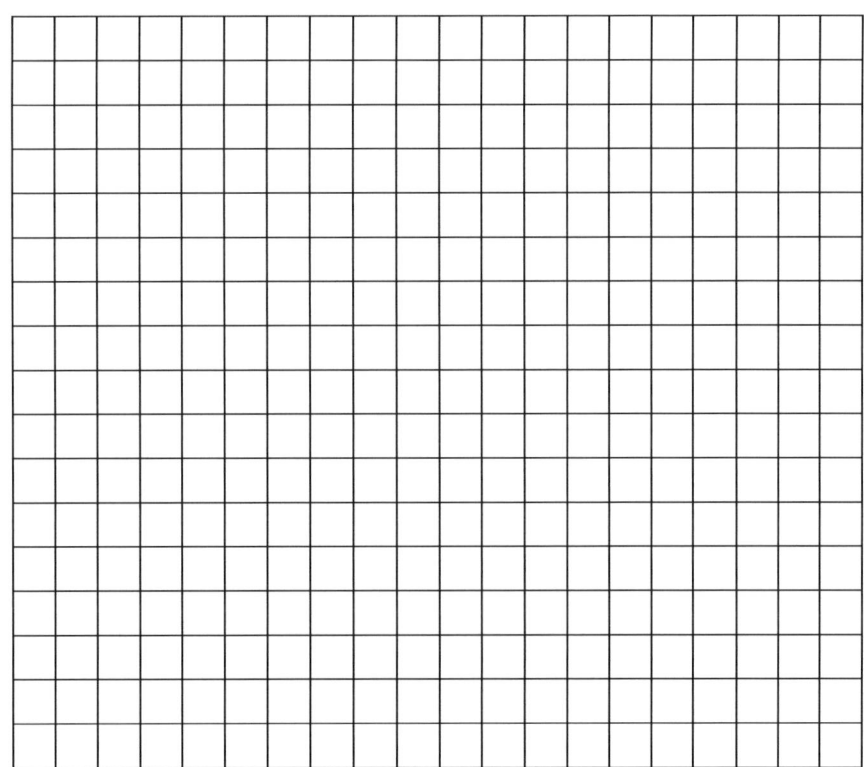

(4 marks)

9. Simply the following fully

i) $\dfrac{24x^5y^2}{49x^{-2}y^3} \times \dfrac{35x^7y^2}{36x^3y^{10}}$

(5 marks)

ii) $\left(\dfrac{7a^3}{5c^2}\right)^3$

(3 marks)
(total 8 marks)

10. A bag has x marbles. 3 of them are red and the rest are black. Georgina took two marbles out without replacing one after the other. The probability of her picking 2 red marbles is $\dfrac{1}{5}$. Work out the value of x.

(6 marks)

11. Solve the following inequality
$$3x^2 - 2x - 16 < 0$$

(5 marks)

12. Show that $y = 3x^2 + 6x + 7$.

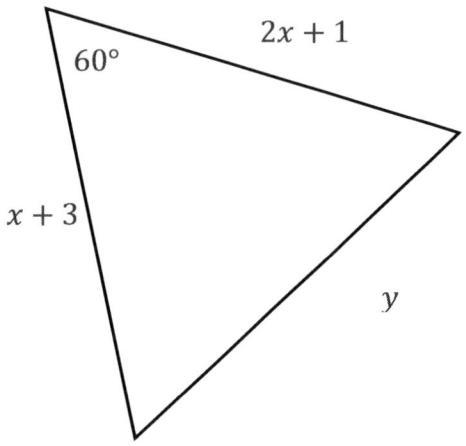

(5 marks)

13. Two cylinders P & Q are mathematically similar to each other. The surface area of cylinder P is $336 cm^2$ and the surface area of cylinder Q is $756 cm^2$. The volume of cylinder Q is $1350 cm^3$. Work out the volume of cylinder P.

(4 marks)

14.

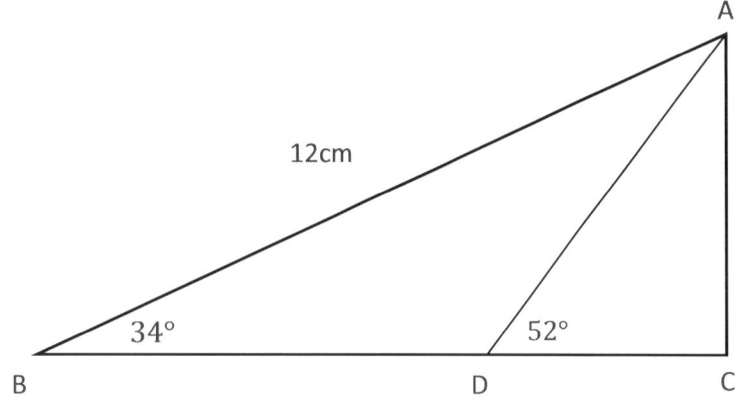

i) Calculate the length BD.

(4 marks)

ii) Calculate the area of the triangle ACD.

(2 marks)
(total 6 marks)

15. Work out the angles DBC, ABD, ACD & DOC. (write reasons for each stage of your working.)

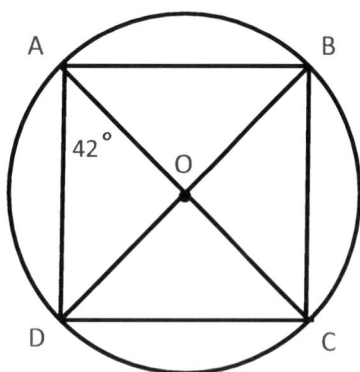

(6 marks)

16. A rectangle has an area of $230cm^2$ correct to 2 significant figures. The length of the rectangle is $30cm$ correct to 1 significant figure.

Calculate the lower bound for the width of the rectangle.

(5 marks)

17. The cuboid ABCDEFGH is drawn below.

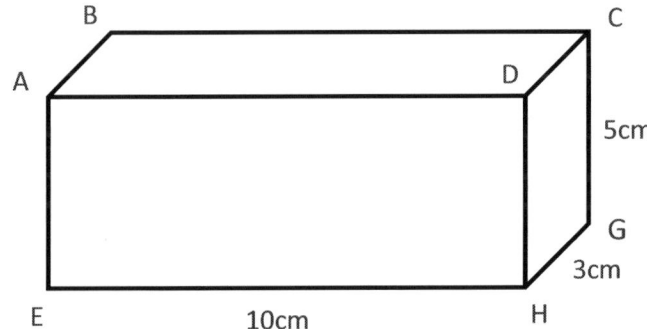

Calculate the angle between the plane ABCD and the line EC.

(6 marks)

18. Prove that the difference between any two odd square numbers is always divisible by 4.

(5 marks)

19. The vectors $\overrightarrow{PT} = 8t$ & $\overrightarrow{PS} = 5s$. The lines TS & RQ are parallel. $PR:RT = 3:5$ & $PQ:QS = 2:3$.

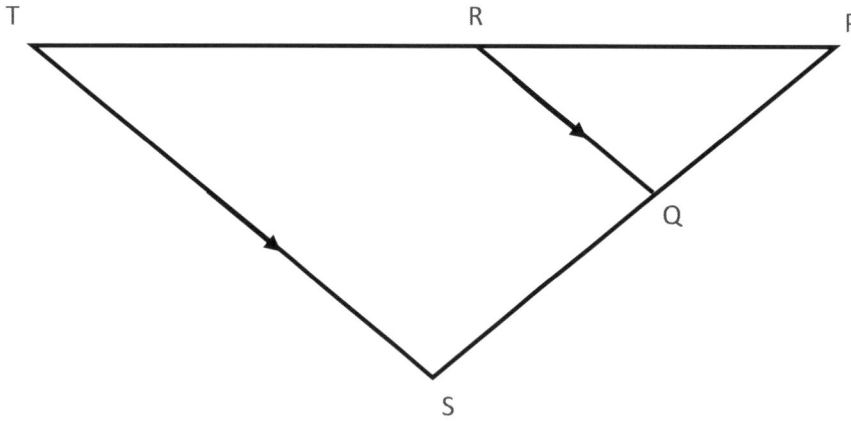

Work out the ratio $RQ:TS$.

(6 marks)

Total for paper: 100 marks

End

Printed in Great Britain
by Amazon